Ladybird umbrella, Sophia, age 5

Little Inventors WANTED

We live in a world that is constantly changing: each day scientists and designers are creating amazing things that once were thought of as being impossible.

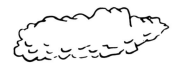

This is where you come in. You see, we believe that children simply have the best imaginations and can think without limits. We **know** children are the inventors of tomorrow.

This book is **YOUR personal invitation** to put that brilliant brain of yours to good use, build up your inventing expertise and start thinking up some wondrous ideas!

Welcome to the Little Inventors handbook, where you will find out how to come up with ideas, draw them and share them with the world!

Chapter 1

For all your ingenious ideas...

Introducing Little Inventors

What is Little Inventors?

We are a team who believe children have the best ideas!

You share your inventions with us, we choose the most ingenious ones and then we ask skilled Magnificent Makers to turn them into real prototypes.

They are shown online but also in exhibitions and museums across the world!

We are looking for Little Inventors to help change the world they live in by solving problems, big or small. This can be as simple as making people smile more, or as complicated as exploring a new planet.

Throughout this book you will learn how to become the most ingenious Little Inventor you can be!

You will find out how to spot problems and different ways to come up with invention ideas. Are you ready?

Little Inventors

See our website at
littleinventors.org

Your Little Inventors passport

My name is

I am

years old

What I really like

What I truly dislike

Things I'd like to do

I want to become an inventor because...

Draw a picture of yourself

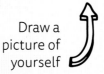

We have Little Inventors all over the world!

 Bone Out

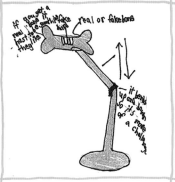

" My invention is called Bone out. It's called bone out because your dog can jump for the treat, and the bone out is made out of metal, plastic hand and a real or fake bone."

Rebeca, age 10
Toronto, Canada

Wiping shoes

" This is a pair of shoes that can help you wipe the floor while walking. It has a lot of duster cloth attached on the bottom of the shoes. It is invented for people who doesn't like to do houseworks. "

Xiaolan, age 14
Guangzhou, China

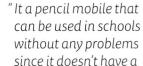

Pencil mobile

" It a pencil mobile that can be used in schools without any problems since it doesn't have a camera or a screen. It have a small number plate and a signal sensor at the top. It can be charged via USB wire. It is easy to use and makes a beeping sound when the battery is low. "

Fatimah, age 10
Qatif, Saudi Arabia

It's time to put **your** name on the map!

Great things can happen when you're a Little Inventor!

Your invention could be shown in exhibitions across the world!

Oliver, 6, invented the High Five Machine!

It is now in the permanent collection at the V&A museum in London!

Your invention could be made real!

Anais, 8, invented the Sweet hat!

It was made real by our very own Chief Inventor, Dominic!

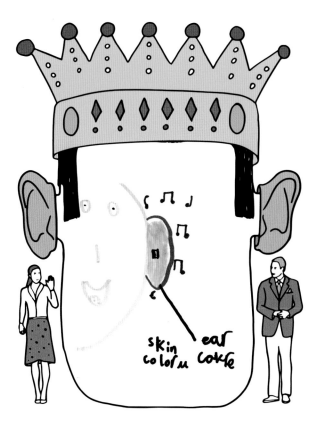

Your idea could even travel to space!

You could meet some really interesting people!

Emily, 8, invented the Silent ear cover.

She got to show and explain her invention to Prince William and Princess Kate!

Our Canadian Little Inventors will see one of their ideas travelling to the International Space Station!

Meet Dominic Wilcox, our Chief Inventor!

Dominic Wilcox is an inventor, designer and artist. His work has been shown in museums and galleries around the world.

He is always coming up with wonderful, intriguing and fun invention ideas!

Photo by Sylvain Deleu

The driverless stained glass car of the future!

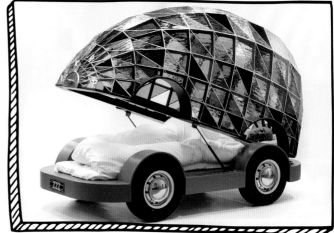

" By 2060, all cars will be driverless and really safe, so I decided to make a glass car with a bed inside, so you can rest while travelling! But it could also have all sorts of other things, like a cafe, a jacuzzi or an office! "

The GPS shoes!

Photo by
Joe McGorty

"I decided to make a pair of shoes that can navigate you to anywhere you wish to travel to. I thought about the 'Wizard of Oz' and how Dorothy could click her heels together to go home.

You enter your postcode and the lights show you which direction to follow and how far you still have to go!"

The tea cup with in-built cooling fan!

"I was thinking that sometimes when my cup of tea is too hot, I can burn my tongue. So I thought maybe I could add a cooling fan onto the saucer."

Dominic is on a mission...

...to inspire children around the world to come up with their own extraordinary, ingenious, or just plain brilliantly bonkers invention ideas.

It's time
to become a
Little Inventor!

Chapter 2

It all starts with a spark...

Why ideas are EVERYTHING

The power of invention!

From inventing the spoon to the latest megarocket, we are constantly thinking about how to make our life easier, better or more fun!

Maybe we're simply daydreaming or chatting with a friend when a thought starts forming in our mind as we ask ourselves...

What if...?

First bridge attempt

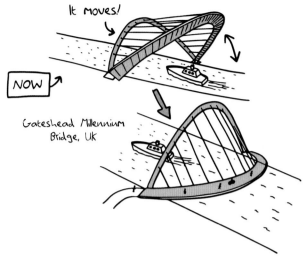

It moves!

NOW

Gateshead Millennium Bridge, UK

100 years ago

200 years ago

1000 years ago

That simple question is the spark that has led us on the exciting journey to the modern world, full of ever more sophisticated inventions and complex technology.

Our future is one where everything is possible, it's just a matter of when...

That is the **magic** of ideas. And everybody can think up ideas!

So where do you start?

We all have little problems in our lives, but we can use invention to solve them. You can find great ideas everywhere, you just need to look closely.

Hot air balloon with side basket

Extractor fan for smells

Waterfall umbrella

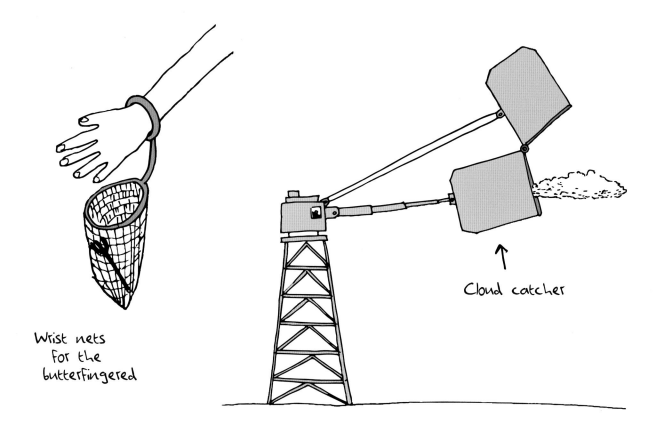

Wrist nets
for the
butterfingered

Cloud catcher

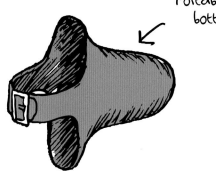

Portable
bottom seat

Mind bubbles!

We have lots of thoughts going through our heads every day. Little thoughts that crop up all the time, that we don't necessarily pay very much attention to.

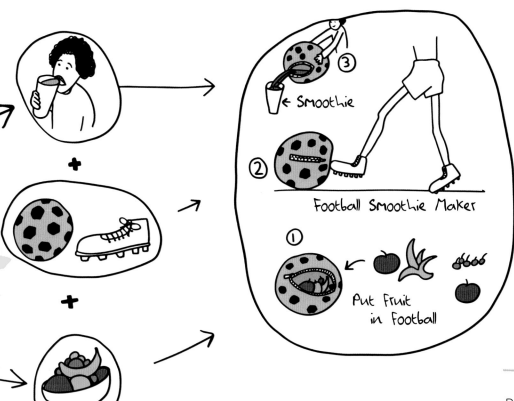

But what happens if you give your thoughts the time and space to bubble up and grow?

This is where ideas are born!

Bring it out of your imagination - and into the real world!

Now that you have an idea tickling your brain, it's time to take the first step to making it real by drawing it!

It might be on a blank piece of paper, a napkin, on sand, it doesn't matter. Jotting your idea down will help grow your idea from a simple thought to a proper invention idea. You don't have to be good at art to make a good invention drawing, the most important thing is to capture your idea.

TOP TIPS:

You might find it easier to start with drawing, writing or naming your invention!

Use labels and colour to really bring it to life.

Find a good catchy name for your invention.

Hook ↗ hat

chief Inventor !

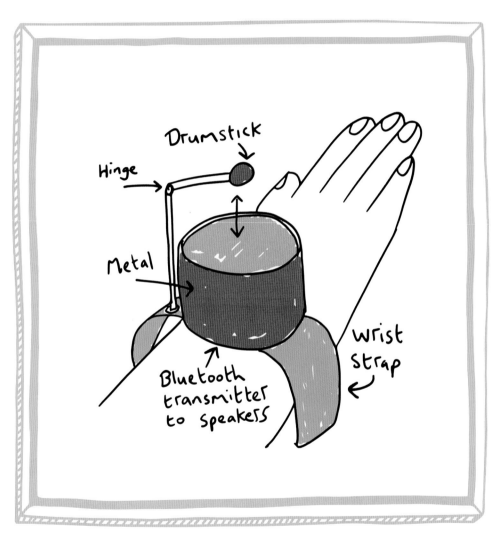

Invention name:
The wearable
wrist drum

By:
Dominic

What it does:
This is a drum that
you can wear on your
wrist, so you can
beat out a rhythm
anywhere!

Your journey to becoming a fully-fledged Little Inventor!

You might already have an invention idea in mind, and that's great! But sometimes it can be hard to get started.

This is what The Little Inventors Handbook is all about: giving you tips and inspiration to keep great ideas flowing!

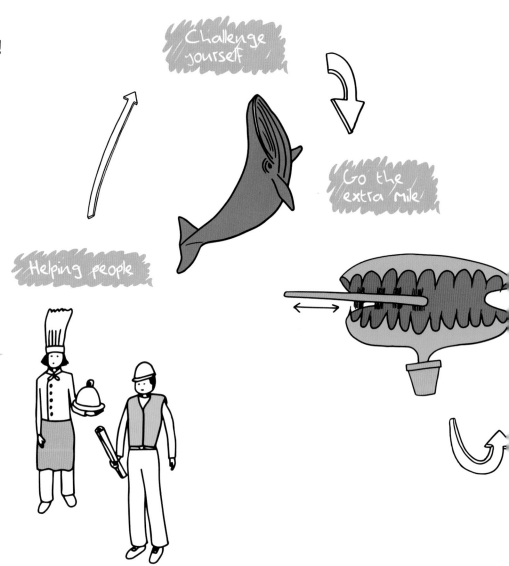

Challenge yourself

Go the extra mile

Helping people

Get making!

The future is yours

You're a Little Inventor!

Keep the ideas coming

It doesn't matter if your idea doesn't seem to be entirely useful or serious or even that GOOD to start with. It's an idea, and the more ideas you have, the more your brain gets used to thinking about solutions to problems.

A lot of inventions were made by mistake, like crisps and fireworks!

One idea might lead to another idea, or it might inspire someone else, or it might get out of your brain to make room for more!

From simple, little ideas, big ideas can **GroW!**

Start simple

You don't need to figure out how to change the world – thinking about small problems or challenges can be just as useful!

What do you find tricky? Is there a simple way to change that?

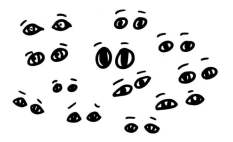

A fresh pair of eyes

Sometimes you just need to step out of your own shoes to see things differently. Once you have a problem, try to imagine what it looks like to someone else.

Break it down

Imagine having to explain your invention to someone who knows nothing about it. How would you explain it?

It can be really helpful to think about things in steps. Break down who the invention is for, what it does and how it works.

Thinking of these steps will help you come up with a better invention.

Reach for the stars

It might seem a little crazy or simply impossible, but so did landing on the moon! So don't be scared to think up a big or bonkers idea – who knows what will come of it?

Often one idea leads to something else that jump-starts your imagination for a brilliant invention!

Now it's time to get inventing!

Chapter 3

A bus driver, a ghost, or a baby...

Helping others

Who needs inventions?

Everyone!

Think about the first humans, who had... well... not very much at all, and now look at the world around you.

Who do you think came up with **chairs, forks, shoes**...? Everything in the world had to be invented at some point!

'Scarf-copter!'

Coming up with inventions is one of the most natural things that humans (and some animals) do. It's how we learn to adapt to our environment, how we solve problems around us.

Everyone here could do with some help...

What sort of invention do you think they might need?

keeps losing balance →

A ballerina

Wants to jump higher

Toes get sore ←

A runner

Wants faster shoes

Needs a drink, but doesn't want to stop ←

keeps losing hat ←

A grandpa

Legs not strong

Finds it difficult to grip things

A Firewoman

Gets very hot near fire

No arms to hold an umbrella!

knocks things over with tail

A dinosaur

Neck gets cold in winter

Likes exercise but doesn't like getting sweaty

A cat

Leaves hair everywhere

A doctor

Too busy

Needs to carry lots of equipment

Needs get to patient more quick!

Feels tired

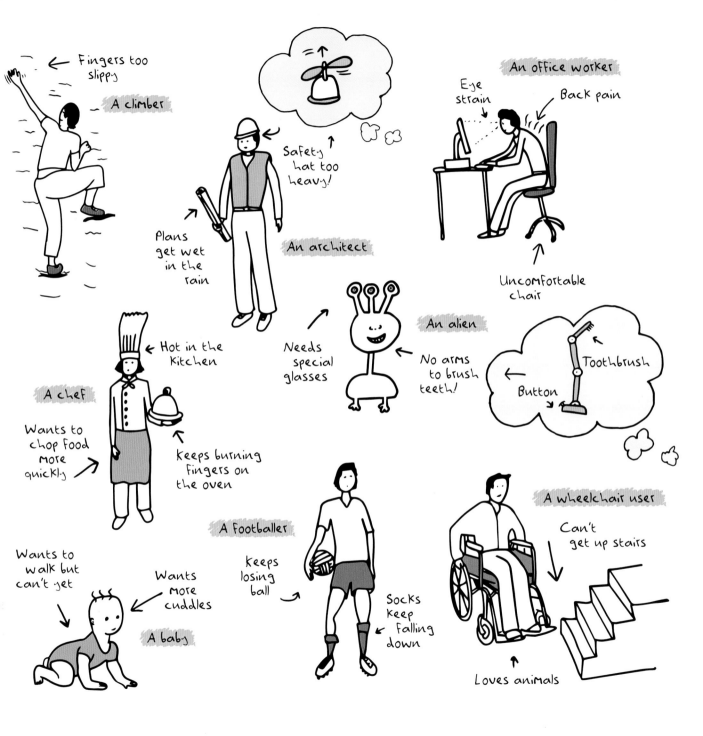

This is Evie.

 She's 1 year old.

 She's a baby!

 Likes: Going on the swings.

 Dislikes: Bus journeys.

 An invention that could help:

A swing-pod bus!

Pod made of glass to see out of

This is Gary.

 He's 403 years old.

 He's a ghost!

 Likes: Scaring people.

 Dislikes: Stains.

An invention that could help:

The Ghost Wash 'n' Dry!

conveyor belt ↓

Who could you help?

It could be someone in your **family** or a **friend**... It could be **you**!

It could be an **astronaut**, a **ballerina** or a **shoemaker**. It could be a **teacher**, a **granny** or an **animal**!

It's up to you.

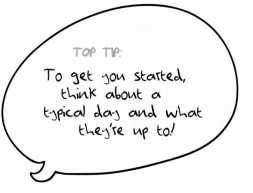

TOP TIP:

To get you started, think about a typical day and what they're up to!

Chief Inventor

↑ Reverse listening device

Now think about their lives, **what they like or dislike, what they find tricky or a bit boring**.

Once you have all of this in mind, it will be much easier to figure out a way to help them!

Create a profile

Draw the person (or thing) your invention is for!

What is their name?

How old are they?

What do they do?

What are their hobbies?

What do they like?

What don't they like?

Your invention idea!

Now it's time for you to come up with your own invention idea!

It could be to help your person (or thing) in a small way or to make their life easier. It could be funny or really bonkers. You're the inventor!

Don't forget to give it a great name and to write down how it works too!

Upload your ideas to **littleinventors.org** for a chance to have it turned into reality!

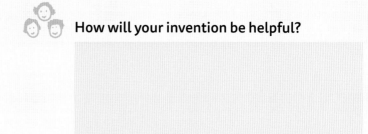

How will your invention be helpful?

How will it work?

My invention:

Name it!

Draw BIG, use colours and add labels!

Well done!

You have thought of your first invention!

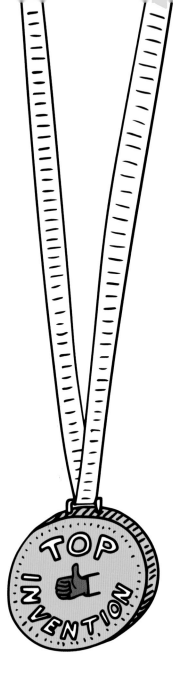

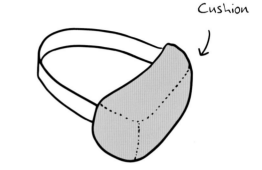

Cushion

The main thing to remember is **inventing is a way to help people**.

If you can understand what makes people's lives a bit difficult or boring, then you are well on your way to finding a way to help them!

There are lots of ways to come up with inventions. The main challenge is to know how to solve the problem...

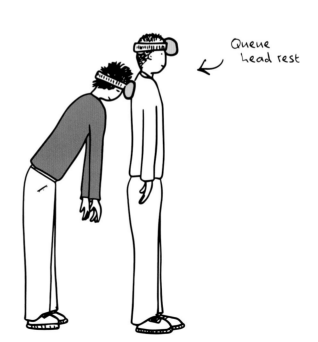

Queue head rest

From the minds of
our Little Inventors...

Ladybird umbrella

Sophia, age 5
Sunderland, UK

Made real by our Magnificent Maker!

Sophia met with expert glass craftsman Norman of Wearside Glass, who came out of retirement to make Sophia's Ladybird umbrella invention at the National Glass Centre, Sunderland.

The idea is simply to give ladybirds a place to keep dry.

What we thought

" Lovely idea Sophia. Caring about bugs getting wet is very nice of you. Well done!"

Chapter 4

Talk about the weather,
pollution or playgrounds...

Rise to the challenge!

Become a challenge spotter!

Sometimes it's hard to think of ideas just out of the blue. But really, it's a lot to do with **being curious** and paying attention to the world around you.

While there have been some incredible inventions over the last few centuries, we still have many big challenges ahead and quite a few small ones too! So keep asking yourself how can we improve the way we do things? How can we find ways to work around big and small issues that are making life difficult or just a little boring?

Time to put your explorer hat on and get personal with the challenges around you and around the world – then start inventing!

Pick a topic!

There are so many challenges in this world (and beyond)! Some are significant and serious issues, some are more fun and lighthearted, but they all have something in common: they are in desperate need of some inventive spirit!

Which of these topics has a challenge that scratches your inventor's itch? It could be one of these, or something else that you add to the list!

Space

Ocean

Food

Animals

Play

Travel

Health

Entertainment

Weather

Wearables

Cities

Environment

Energy

What's YOUR topic?

Once you have your topic, you can explore it!

Starting with a big topic gives us the freedom to be really creative. To kick-start our imaginations we begin by thinking of words around the subject, and potential challenges we could make solutions for.

Here is an example for the ocean with just some ideas that came into our heads – it helps to approach it from lots of different angles!

Rowing

Cruise

Straws

Shopping bags

Pirates

POLLUTION

BOAT

Plastic

Whales

Coral reef

SEA LIFE

Sunglasses

Fishing

The Ocean

BEACH

Suntan cream

Sand-castles

WATER

Waves

Deep

SWIMMING

Salty

Danger

Swimsuit

Armbands

Now where could these words take us when we play with them? Some invention ideas could be:

Inflatable swim-suit for nervous swimmers

Plastivore whale cruise ship

- Need confidence to swim
- Everyone wears a swimsuit
- Inflatable for safety
- Need to remove plastic
- Whales filter their food
- Cruise ships cross the oceans

Your challenge idea!

Now it's your turn, so prepare those brain cogs for turning!

Start by writing your topic in the centre of the page, and then add words around it that come to you when thinking about your topic and challenge.

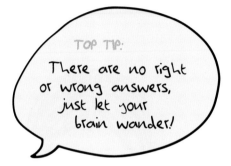

TOP TIP:
There are no right or wrong answers, just let your brain wander!

↑
Propeller hat

Chief Inventor

You can use colours, shapes and drawings too, whatever goes through your mind – this is why this technique is called a mind map!

It's all about building a bigger picture and expanding your thoughts in lots of different directions – hopefully finding one that you can choose to take forward.

Your mind map

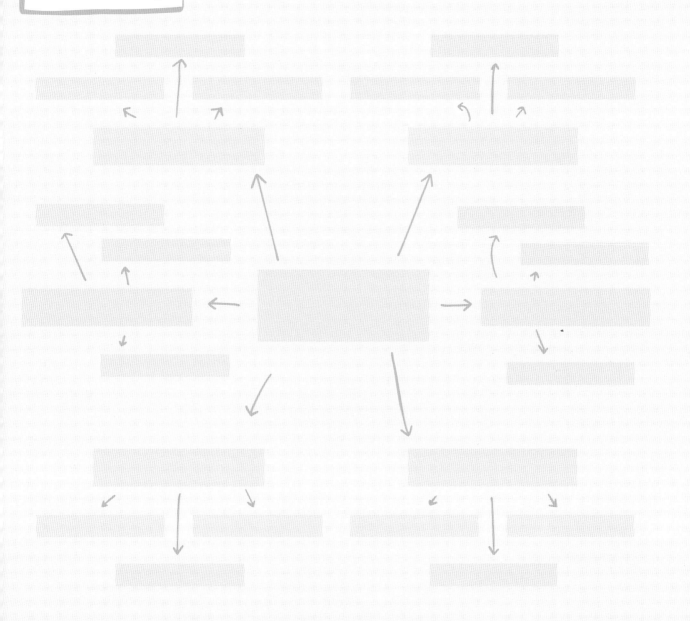

Time to draw your invention

Now that you have created your mind map and thought about your topic, the challenges within it, and maybe some fun ideas – it is time to switch to full-on inventor mode!

Using your mind map as inspiration, come up with an invention idea to either solve a problem, make something fun or change the way we live.

Upload your ideas to **littleinventors.org** for a chance to have it turned into reality!

What is the challenge you are tackling?

What are the key words that inspire you?

What makes your invention interesting?

My invention:

Name it!

Draw BIG, use colours and add labels!

Well done!

You have taken up your first challenge!

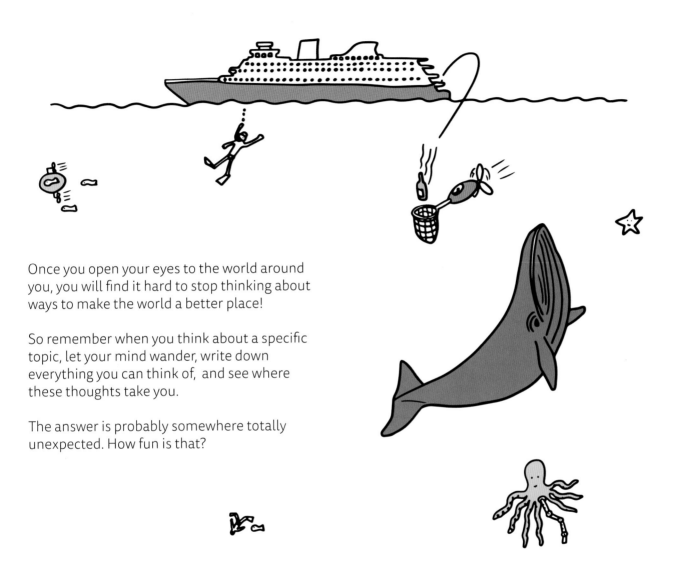

Once you open your eyes to the world around you, you will find it hard to stop thinking about ways to make the world a better place!

So remember when you think about a specific topic, let your mind wander, write down everything you can think of, and see where these thoughts take you.

The answer is probably somewhere totally unexpected. How fun is that?

From the minds of our Little Inventors...

No pollution jacket

Gruff, age 6
Topic: Car pollution!

" This is for people who want to keep away from the car fumes and people with asthma like me.

It keeps car fumes away from the person wearing it. It is made of waterproof fabric (polyester) and has a filter over the face that you can see through.

The jacket sleeves are reflective so people can see you."

Pollution filter

Reflective sleeves

legs

Made real by our Magnificent Maker!

Gruff's idea was made real by textile designer Barley Massey in her shop 'Fabrications' in London.

Once Barley had made the jacket and trousers, we took them to Zoe Laughlin, Director of The Institute of Making in London. She added a little bit of science to Gruff's jacket by spraying it with titanium dioxide. This substance is known to break down pollutants in the atmosphere!

What we thought

" Pollution is dark and grey so we love how you want to wear a bright piece of clothing to filter the pollution away."

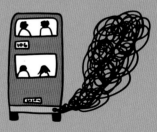

Can you imagine how YOUR invention could change the world?

Chapter 5

Give your boots a reboot,
add a spring to your step!

Add extra
to the
ordinary

From bicycles to telephones...

...vacuum cleaners to light bulbs, we tend to take the objects that we use every day for granted.

Yet a lot of these started life as a simple solution to a simple problem. They then evolved through the years or even centuries, and were modified to be more efficient or adapted to use new tricks of technology.

A brilliant way to start inventing is to take an existing object and imagining new ways to improve it or use it for something different.

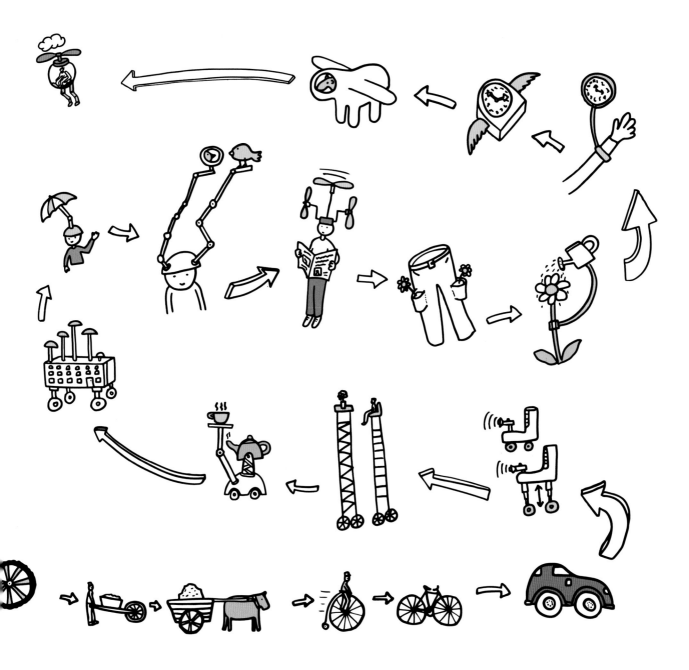

Reinventing normal!

Reinventing normal is our Chief Inventor's favourite method of inventing! It's a great way to find fun in everyday objects, give them a different purpose and push the limits of their use.

It may be as simple as putting two objects together that would not normally meet in real life and seeing what the result is!

Why don't you have a go?

Electric

Dodgem roller skates

Bunk chair

Toothbrush maracas

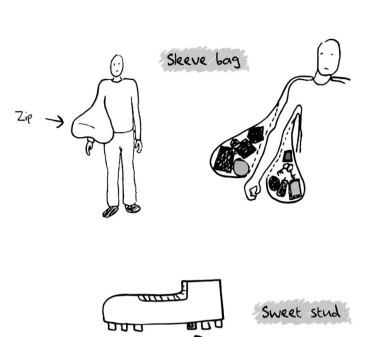

Sleeve bag

Zip →

Door book

Sweet stud

Secret place
for a sweet

Slide chair

TOP TIP:

A good way to start is
to really think about one
object and explore all its
features before figuring
out how it could
be made extraordinary...

The toothbrush!

Think of yourself as an object detective!

Your mission?
Uncover as much information as you can about a specific everyday object.

What happens if you ask yourself the following questions about the humble toothbrush, for example?

Soft bristles

Flexible

Thumb grip

Finger grip

Who can it be used by?

Anyone with teeth
(humans, rabbits, Venus flytraps)

What is it for?

To remove food bits from teeth and gums and keep mouths healthy

What are the main features and why is it shaped that way?

- The handle helps by holding the brushes
- It's long to reach the back teeth
- The bristles are soft so they don't hurt

Are there other objects in the same family?

Toilet brush, broom, paint brush, toothpaste, dentures

What else could it be used for?

Brush something small or hard to reach

Can you name some other objects that have nothing to do with it?

Hat, frying pan, violin, desk lamp, umbrella

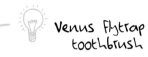 Venus flytrap
toothbrush

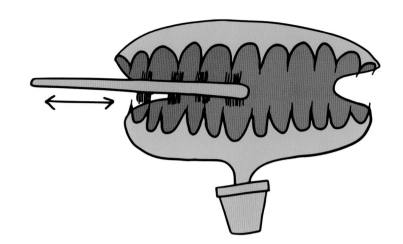

 Hamster tickling device

 Toothbrush
violin bow

 Toothbrush
umbrella

Frying pan
toothbrush

Get on the case!

Now it's your turn: Draw an everyday object to investigate why it is the way it is, and what clues its features might reveal.

Who can it be used by?

What is it for?

What are the main features and why is it shaped that way?

-
-
-

Are there other objects in the same family?

What else could it be used for?

Can you name some other objects that have nothing to do with it?

Now, can you use what you have found out to think of a way to improve your object?

Can you add something to it, maybe a new feature such as a different shape, another use, or an extra part?

My new feature is...

Maybe you can take inspiration from another object and bring them together...

What's the other object?

How do they fit together?

Why does it make this a new, better or more fun object?

Bring it all together

Now that you have thought about what specific objects are designed for and thought about ways you can transform them, it's time to get your very own EXTRA-ordinary invention going!

You can use the object you have already thought about, or a totally new one, it's up to you. After all, YOU are the inventor!

Upload your ideas to **littleinventors.org** for a chance to have it turned into reality!

What is the object that you are starting with?

What do you think could improve it?

Who will use this new object and why will they love it?

My invention:

Name it!

Draw BIG, use colours and add labels!

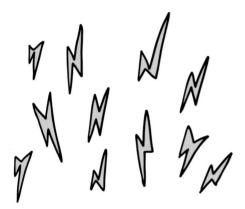

Woohoo!

Now that's taking things up a notch!

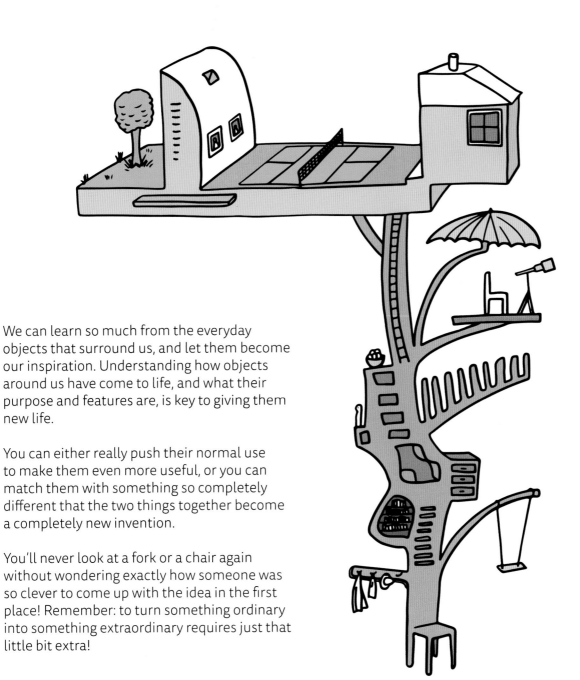

We can learn so much from the everyday objects that surround us, and let them become our inspiration. Understanding how objects around us have come to life, and what their purpose and features are, is key to giving them new life.

You can either really push their normal use to make them even more useful, or you can match them with something so completely different that the two things together become a completely new invention.

You'll never look at a fork or a chair again without wondering exactly how someone was so clever to come up with the idea in the first place! Remember: to turn something ordinary into something extraordinary requires just that little bit extra!

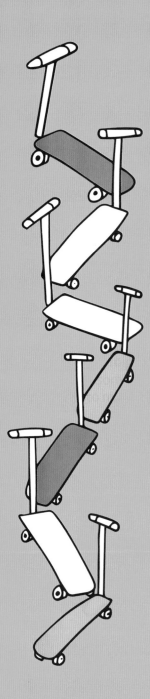

From the minds of our Little Inventors...

Family scooter

Wendy, age 9
Sunderland, UK

" This invention is a family scooter. It works by all the family push and it rides. It would be great for a big family. "

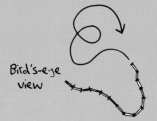

Bird's-eye view

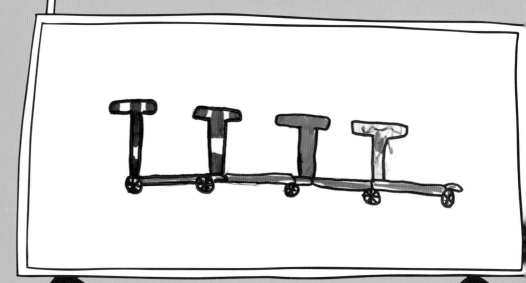

Made real by our Magnificent Maker!

Wendy's idea was made real by Roger O'Brien at the faculty of Engineering at the University of Sunderland.

" The team at the Institute for Automotive & Manufacturing Advanced Practice thought the Family scooter concept was a great idea for their Formula Student race team to enjoy and get involved with. It is fun and brilliant in its simplicity!

Wendy's invention sketch was clear and conveyed much of the information we needed. We met with Wendy at AMAP so she could get involved in assembling the final version of the scooter."

Chapter 6

Wacky, zany
or seriously silly...

Go truly
bonkers

Silly is mighty!

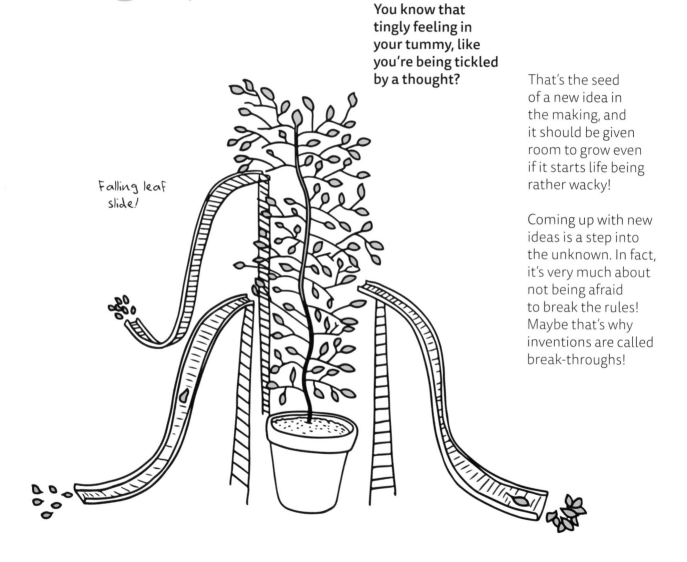

Falling leaf slide!

You know that tingly feeling in your tummy, like you're being tickled by a thought?

That's the seed of a new idea in the making, and it should be given room to grow even if it starts life being rather wacky!

Coming up with new ideas is a step into the unknown. In fact, it's very much about not being afraid to break the rules! Maybe that's why inventions are called break-throughs!

To get going, you have to allow and encourage your mind to make leaps big or small, and connect your thoughts without judgement. Who knows where this spark will lead – if it gets your brain ticking, it's already working!

No two people's imaginations are the same, and no one else can think like you, so embrace the silly, wacky and zany and let your mind take you where no one else has been before!

Yo-yo bungee!

Playtime for your mind!

The best way to learn to think outside the box is to allow your mind to roam freely and follow its whims – indulge your imagination!

It's working out pretty well for our Chief Inventor, who is a master of thought-provoking and playful ideas...

Walk forwards to move backwards

Salted thumb lolly

Sandwich making crane

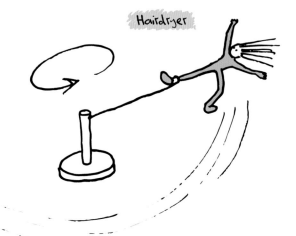

Hairdryer

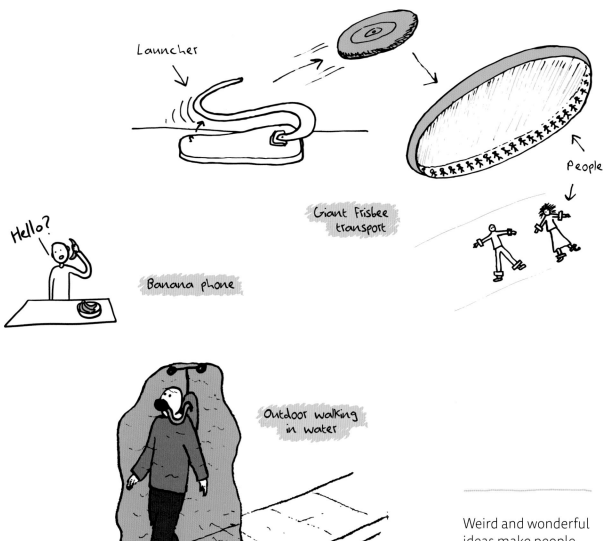

Launcher

Giant Frisbee transport

People

Hello?

Banana phone

Outdoor walking in water

Weird and wonderful ideas make people smile. Smiling and having happy thoughts are good for our minds!

Coming up with new ideas is about allowing leaps of imagination!

Here are some tips on how to kick-start your thinking and be truly...

Opposites

Putting two things that don't belong together can bring up some surprises and really move your thinking out of the comfort zone.

Bold

Be brave, be bold, don't worry what others think. Now is not the time to be shy! You are going where no one has gone before!

No!

Just say a firm 'no' to your inner critic.

Forget about all the rules and the dos and don'ts, what is possible or what isn't. It's your time to open the door to all ideas!

Kaleidoscope

Look at things from all different angles and mix it up!

Break down how things look and feel and discover them under a new light!

Extra-wacky!

Imagine being a crab, a tree or even a slice of pizza – how do you see the world?

Ask yourself what your main qualities are, and how you can apply them to something else.

What would you want if you were a bike?

Random

Playing freely with random words helps get access to all the information that is already stored in your brain. Pick any word and see what your brain picks next!

Stretch

Think about extremes: what if something was very big or very small, very long or very soft, or if you had a thousand instead of one.

Now it's your chance to really go all out!

Random words string theory!

Food:

Animal:

Place:

Object:

Colour:

Activity:

Write down the first words that pop into your head.

Do it fast and without stopping, or you might start thinking about it too much!

Now string all of these words together in a sentence that tells a story:

Meet your inner frog

Write the first word that you can think of on the first lily pad – then allow your mind to jump to the next leaf. It could be a world away from your first word. Again, don't try to think too much about what it means, let your frog mind leap happily!

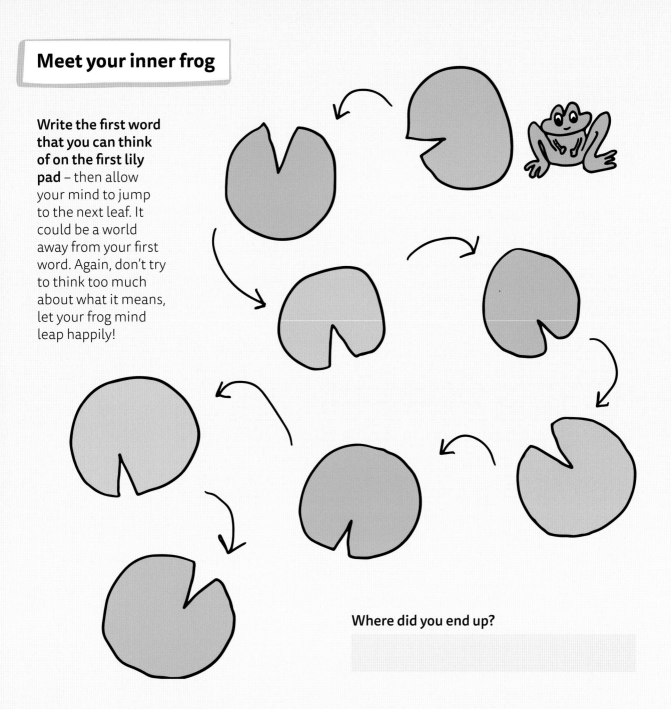

Where did you end up?

How zany can you get?

Now that you have practised thinking in a more random way are you ready to think up a topsy-turvy invention?

You know what we are going to say here: there are no rules, it's just you and that fabulous brain of yours...

You can use this sticky note to draw/make notes, whatever works for you. Flick back to the words you came up with on the last page to help keep your mind open!

Upload your ideas to **littleinventors.org** for a chance to have it turned into reality!

My invention:

Name it!

Draw BIG, use colours and add labels!

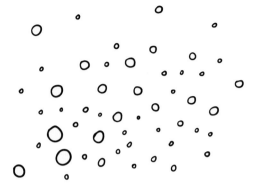

Aaaaaah!

Doesn't it feel great to really let loose?

Bet you really didn't expect what you came up with!

The thing with ideas is that they are like little bursts of surprises. Your brain is the most amazing computer that can crunch thoughts into your very own unique invention ideas – and they don't need to be serious or realistic. The simple seed of a good idea can come from anywhere.

Sometimes it's best to let your mind work its magic, making leaps and connections between words or thoughts by itself, without worrying about it or questioning the result.

You're giving your brain exercise and training to come up with more and more exciting ideas. This is why creativity should be free from the normal rules, and ideas can be anything you think of!

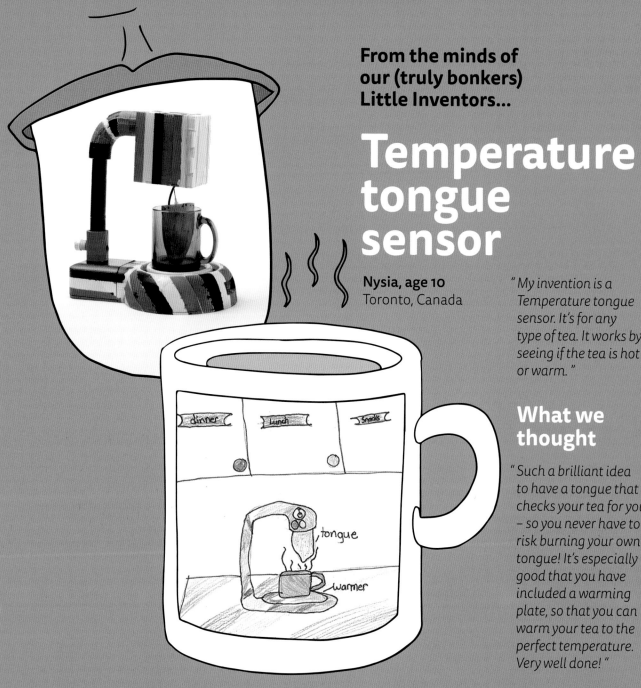

From the minds of our (truly bonkers) Little Inventors...

Temperature tongue sensor

Nysia, age 10
Toronto, Canada

" My invention is a Temperature tongue sensor. It's for any type of tea. It works by seeing if the tea is hot or warm. "

What we thought

" Such a brilliant idea to have a tongue that checks your tea for you – so you never have to risk burning your own tongue! It's especially good that you have included a warming plate, so that you can warm your tea to the perfect temperature. Very well done! "

dinner   lunch   snacks

tongue

warmer

Made real by our Magnificent Maker!

Nysia's idea was made real by Meera Balendran and Kyle Myers from STEAMLabs in Toronto.

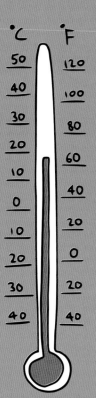

TOP TIP:

It helps to talk about your ideas with a friend

" The minute we saw this invention we knew we had to make it. The combination of its silliness, crazy colour scheme, and practicality made us fall in love with it! We wanted this invention to be as close as possible to Nysia's illustration and description, so we spent a lot of time thinking about what we could make it out of. We decided to 3D print the entire body including the tongue, which is made from flexible filament. We also thought it'd be a nice touch to add in a servo to make the tongue flick around like a real tongue."

Chapter 7

A giant step for humanity
(and dogs?)

Here's
to the
future

What's in the future?

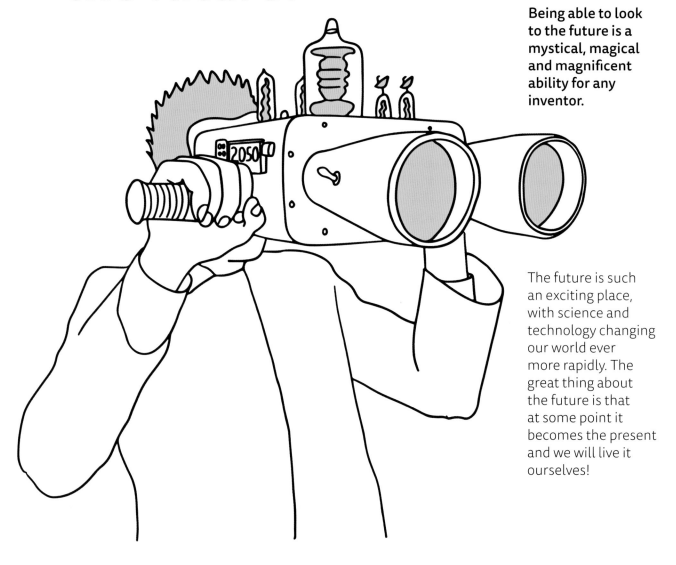

Being able to look to the future is a mystical, magical and magnificent ability for any inventor.

The future is such an exciting place, with science and technology changing our world ever more rapidly. The great thing about the future is that at some point it becomes the present and we will live it ourselves!

This is when science fiction becomes real. There is a good chance that in the future machines will do a lot of our work for us (they already do quite a lot of it) and we will find ourselves with a lot more free time. How do you think we should use it?

There will also be loads more people on Earth – some predict 10 billion by 2050! This will bring some big challenges and we need to keep up!

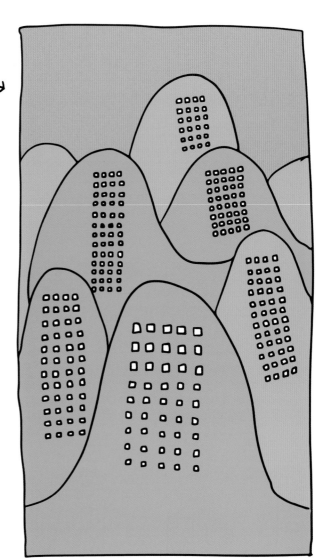

Mountain homes

Imagine our future!

Advances in science could enable us to understand our world much better.

This may mean we might...

Consult robot doctors

Have sky farms

3D print everything we need

All really exciting things! But can you think of the challenges we might face in the future?

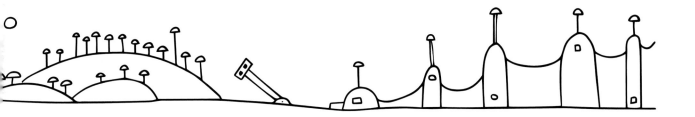

Take a peek at tomorrow's world!

One of the biggest challenges in the future will be overpopulation: as people get healthier and live longer, our planet remains the same size.

How will this affect us? This is when we can project ourselves forward and try to make some predictions about what life might be like in five, ten or twenty years time!

More people means more pollution

How do we deal with the bad things we create, like too much rubbish, CO_2 and plastic...?

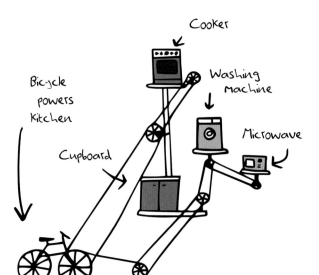

We will need to feed everyone

How about new types of food, new ways of cooking...?

Preserve the natural world

What about keeping green places, looking after wildlife and the oceans?

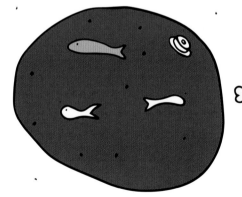

An animal reserve in space?

Could you live in a sky capsule?

There might not be enough land to live on.

Explore different types of living: in the sea, the sky, underground, in orbit or on other planets.

Meet the future you!

What will matter for the whole of humanity will also affect you directly...

Let's light leap twenty years from now, can you predict what your future life could be like?

Draw your future self

Where do you live?

How do you get to where you need to go?

What job do you have?

What do you do for fun?

What three things didn't exist when you were little?

Think about things in your home, your bag, what you wear or how you do things!

1

2

3

How is your world different now?

What do you find challenging every day?

What is your invention for the future?

Now that you are travelling through time to get a glimpse of all the challenges and possibilities that lie ahead of us, you can start thinking about an invention idea that would fit in perfectly!

Upload your ideas to **littleinventors.org** for a chance to have it turned into reality!

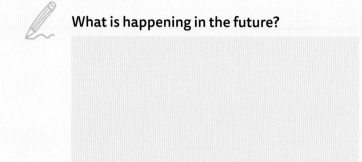

What is happening in the future?

What is the challenge you want to solve?

My invention:

Name it!

Draw BIG, use colours and add labels!

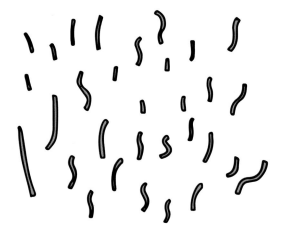

Cosmic!

**How did you like
peering into the future?**

Little

Inventors

The one thing that machines can't quite replace is **our inventive spirit**.

We will always need creative people LIKE YOU to come up with invention ideas today to make sure we have a better tomorrow.

From the minds of our Little Inventors...

The smart house

Adriana, age 10
Newcastle upon Tyne, UK

What we thought

"From the shape of your house to your invented super glass and your staircase ramps, there is much to love and admire about your very smart house, Adriana - that's a lot of inventions in one!"

Made real by our Magnificent Maker!

Adriana's idea made real by FaulknerBrowns Architects, Newcastle upon Tyne, UK.

FaulknerBrowns architect Nathalie Baxter said:

"It has been a pleasure to help Adriana develop her invention from a two-dimensional sectional drawing into a three-dimensional design solution. Her attention to detail is very impressive. Adriana has incorporated a series of small-scale inventions such as rainwater harvesting, waste recycling, photovoltaic glazing and water recycling to make sustainable living a convenient choice in this home."

Photos by FaulknerBrowns!

Chapter 8

Ordinary toilet paper tube or crafty telescope?

Get making!

From 2D to 3D!

Drawing is a fantastic way to capture your ideas when inventing and should always be the first step. Now we need to propel them into the next dimension – off the paper and into reality – from 2D to 3D!

The first step to a real invention is to make a model of it. This is called prototyping. It doesn't need to actually 'work', it's really to start seeing how your object could look as a real thing.

Designers, architects and artists all do it. They get messy with materials to feel what their idea is like in their hands before they embark on a project, and spend quite a long time making models to make sure they get it right.

It's a lot of fun! So let's get making!

Making prototypes doesn't need special materials or skills – you can find lots of things around the house that you could use (but make sure it's OK with a grown-up!). For example, cardboard is great, it's everywhere and you can cut it, roll it, squish it and fold it into pretty much anything!

There is no right or wrong way, it's all about feeling your object in your hands and seeing if it looks right. If it doesn't you can easily change it and try again.

You could use cardboard packaging, old shoe boxes, paper plates or paper cups of different textures and thickness.

Keep empty cereal boxes, egg boxes or toilet tubes and you will have some ready-made shapes to play with. What a fun way to recycle!

These are just some of the things you could use, but the list doesn't have to stop there. It's another chance for you to use your imagination!

Thinking in 3D...

Once you have your invention drawing, it's a good idea to **think about how it will look from different sides.** This will help you to start imagining what it will look like as a real object.

Chief Inventor Dominic invented **the snack shoe**!

" Sometimes you get hungry but there are no shops around. This way I can always have a quick snack when I need it. "

Side view

Your snack

Snack door

Robotic arm

Your snack

Open snack door

Bird's-eye view

Back view

Robotic arm

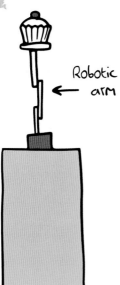

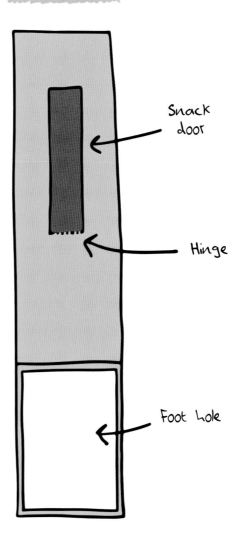

Snack door

Hinge

Foot hole

Prototyping the snack shoe!

Before he started, Dominic had to think about how big his invention was going to be. He wanted it to be a life-size version!

1. Draw the outline of the side of the shoes.

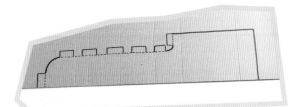

2. Cut them up!

TOP TIP:

You can see me making the snack shoe at littleinventors.org

chief Inventor

3. Use the first shape as a guide to draw and cut all the parts you need.

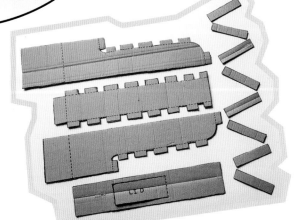

6. Use scissors to make holes and tie all the arm pieces together with string.

7. Tape to the inside of the shoe.

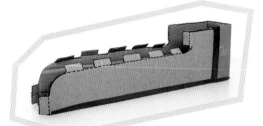

5. Add tape to secure the parts, and for decoration!

You're done!

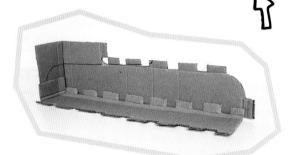

4. Glue the parts together.

My invention:

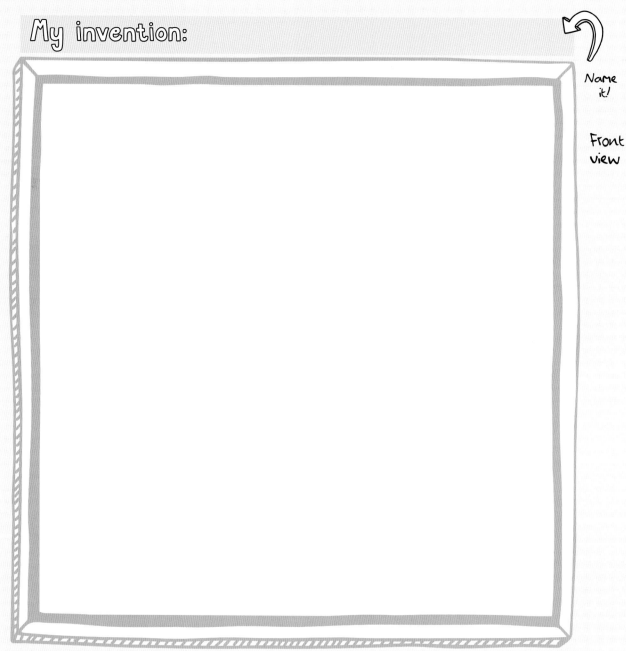

Name it!

Front view

Draw BIG, use colours and add labels!

Now it's your turn...
Plan your prototype!

Bird's-eye view

Side view

Back view

Now get making!

Think about:

- The size of your prototype

- The shapes and elements that make up your invention

- The way they connect together

Then it's really about **getting started and having fun!**

Some fun techniques you could try:

Layering

Slotting

Bending

Texturing

By creating a 3D model of an invention idea **we can start to make more decisions about its design**.

We can see if it should be bigger or smaller, more angular or soft. We can find out if our idea on paper works in reality, or if it needs to be altered to work better.

What have you learned about your invention?

Wow!

It's alive!!!

Having a go at making your idea into a 3D object can seem daunting but it will also help you continue thinking about your idea.

You will figure out what works best and what works less well. And that might make you think about how you can improve on your original idea!

Remember: everything you do is helping you to understand your idea better, so you can take it even further – and who knows, maybe even make it into a real working prototype too!

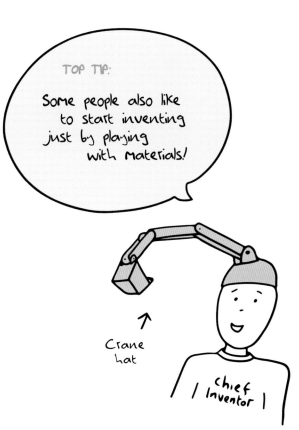

TOP TIP:

Some people also like to start inventing just by playing with materials!

Crane hat

chief Inventor

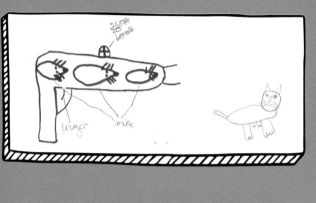

The super toy mouse shooter

What we thought

" Our team have lots of cats and know they would absolutely LOVE this mouse shooter! Well done Ben, great inventing – keep the ideas coming! "

Ben, age 10
Billingham, UK

" My invention shoots soft mice out. It is for my cat to entertain it. Also it is made out of plastic. Buy it for your cat now 10% off. "

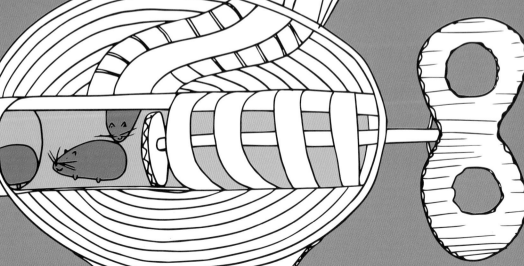

Made real by our Magnificent Maker!

Ben's idea made real by Chloe Rodham, Animator from Newcastle upon Tyne.

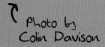

Photo by Colin Davison

"I liked it because I thought it was definitely something my own cat would enjoy and the idea made me smile. I also liked the entrepreneurial thinking behind the design, which included an introductory offer of 10% off.

This design was also appropriate to construct using only cardboard, glue and tape. Because we were using cardboard, the construction was easy to adjust and rework as I built it. I had to think about how the device would function practically, using the illustrated design as a starting point. Making the prototype was so much fun!"

What about your prototype? How different is it to your drawing? Would you change anything?

Chapter 9

Ideas aplenty,
inventions at the ready...

I'm a Little Inventor!

You did it!

Congratulations!

**You have now completed the
Little Inventors handbook!**

**You are officially a Little
Inventor – ready to head out
into the world, and prepared
for anything!**

Share your ideas with the world!

We've shown you how to come up with and develop your ideas, different ways of inspiring your imagination, how to start modelling them and even how to explore them further. It's now time to share your ideas with us – and the rest of the world!

Every time you see this icon in your book, take a picture of your invention and upload it to our website **littleinventors.org**

You can follow the easy steps online (maybe with help from a grown-up).

We look at ALL the drawings that come through and we can't wait to see what you have come up with!

As you have seen throughout this book, some of the invention ideas uploaded are picked to be made into a real object by one of our Magnificent Makers... imagine that!

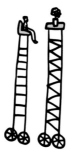

Once you have uploaded, you will receive your very own **certificate** too!

And as your invention will be visible online, you and your family will be able to share your ideas with others too.

Little Inventors

This
Certificate
is awarded to

Little Inventor
Emily

· · ·

For the
Ingenious idea
Silent ear cover

skin colour   ear cover

Dominic Wilcox
Dominic Wilcox
Chief Inventor

But, wait, that's not all! There's always more you can do!

Taking your invention further!

You had a brilliant invention idea, you've drawn it, you might even have made a model of it. But that doesn't mean your invention journey stops there.

Our imagination doesn't stand still and inventions are the same – they can always be tweaked, improved or totally reinvented!

TOP TIP:

Going back to an invention later will help you make it better or simply see a different side to it!

Birdhouse hat

Chief Inventor

Take a step back

When we write or draw or invent, we get involved in detail. Once you are done, it's good to step back and look at the heart of your idea again with a fresh mind.

You might pick up on things you hadn't thought about or other details you want to change or add.

It's a bit like adding a layer of thinking to your invention!

Do some research

You came up with your invention for a specific purpose or person.

Widen your horizons by talking to other people and asking their opinion about your idea. You might be surprised by their reaction, and maybe you will find unexpected uses for it in the process!

Expand your horizon

Your idea will help with one specific challenge or topic. You could use the mind map technique (in chapter 4) to see what other ideas come to you, using your invention as the starting point.

It might get you to think about your invention in a different way or even come up with a whole new invention, related or not!

Draw it again!

Let a bit of time pass and have a go at re-drawing your invention from memory.

How different is it? What do you want to change? Is your idea the same or has it moved on a bit? Just let yourself go with the flow and draw what comes to mind.

Once upon an invention...

Having an invention idea is great, but now can you think what happens next?
This is your chance to write a short story about your invention idea.

Think about...

Where and when it takes place:

The characters in your story:

How they use the invention:

What happens as a result:

How it makes a difference:

My story about...

Invention name:

Write your story

What do you think about your invention now?
You might discover a new use for your invention, or decide to change it or make another one!

Inventors log

Sometimes we have more ideas than we can handle in one go. You can't always draw ideas straight away, but you don't want to lose a good idea, do you?

 List of ideas to explore later

When you have a chance, you can draw them on the following drawing sheets, or download more from **littleinventors.org** – where we add new challenges all the time too!

Who is your invention for:

My invention will help because:

What is the challenge:

How it works:

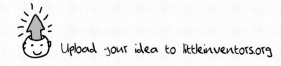

Upload your idea to littleinventors.org

My invention:

Name it!

Draw BIG, use colours and add labels!

Who is your invention for:

My invention will help because:

What is the challenge:

How it works:

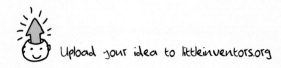

Upload your idea to littleinventors.org

My invention:

Name it!

Draw BIG, use colours and add labels!

Who is your invention for:

My invention will help because:

What is the challenge:

How it works:

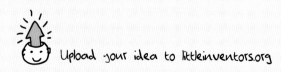

Upload your idea to littleinventors.org

My invention:

Name it!

Draw BIG, use colours and add labels!

Chapter 10

Why inventing is great...

Even if you're a grown-up

Invention can change the world

As you've seen throughout this handbook, inventing is really fun!

But at the heart of it there is also a real purpose. We believe that taking children's ideas seriously can truly make a huge difference to all of us as a society.

Creativity and problem-solving are already among the most sought-after qualities in employees and this is only going to become more important as we head into a future of increased automation and fast-advancing technology.

Using invention as a tool provides children (of all ages) with the freedom to explore science, technology, art and design playfully. It empowers them to believe in the value of their own ideas and ultimately helps them to build up confidence in their own abilities.

Not everyone will become an inventor, but they will all find their inventive spirit of great value when faced with any situation in everyday life or future jobs.

By teaching them to help themselves and other people to come up with solutions, inventing opens children's minds to being more thoughtful. It gives them the right attitude to tackle the challenges ahead as individuals, but also as caring, thoughtful inhabitants of our planet (and beyond!).

We also believe that as adults, we can learn a lot from the unrestricted imagination of children. Just take a few minutes to suspend belief and browse through these ideas, join in their endless curiosity and let yourself be tickled, surprised or intrigued. Thinking like a child is likely to make you reconsider how you view your own world.

After all, it's only by challenging what we think we know that we can bring true change and progress, something that comes very naturally to our wonderful Little Inventors.

Little Inventors top tips for grown-ups.

Here are some simple things you can do to help your Little Inventors feel more confident when coming up with invention ideas.

Put them in the driving seat

It's their invention, their idea, so they can do whatever they want with it!

You know them best

If they are struggling to come up with ideas, think about what they like and what is in their world. Make it personal and relatable.

Nothing is impossible

As they play with ideas, physics and reality are irrelevant, so give them the freedom to explore, with no limits.

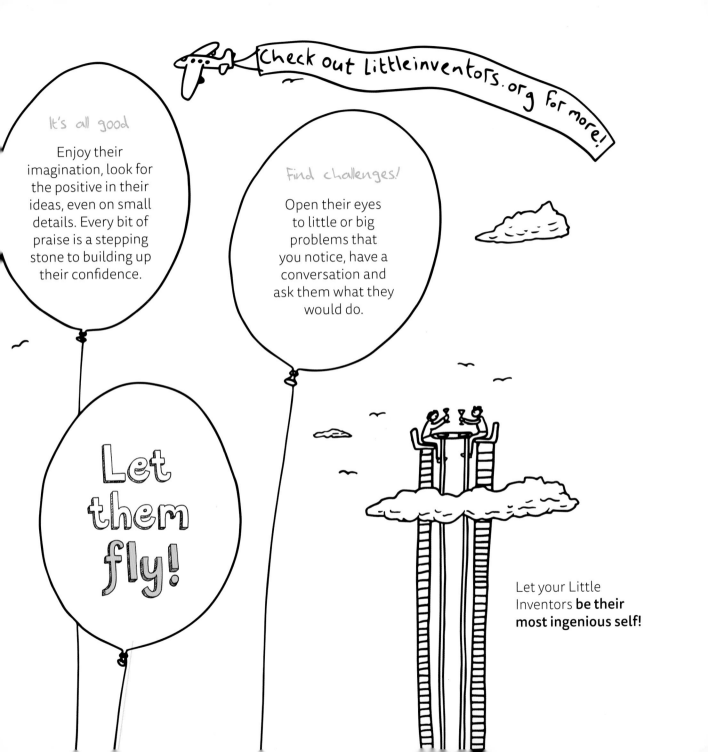

Check out littleinventors.org for more!

It's all good

Enjoy their imagination, look for the positive in their ideas, even on small details. Every bit of praise is a stepping stone to building up their confidence.

Find challenges!

Open their eyes to little or big problems that you notice, have a conversation and ask them what they would do.

Let them fly!

Let your Little Inventors **be their most ingenious self!**

Credits

Ideas, drawings and inspiration by:
Dominic Wilcox

Design by:
Naomi Atkinson

" Why, sometimes I've believed as many as six impossible things before breakfast."

Queen of Hearts,
Alice's Adventures in Wonderland

Written by:
Katherine Mengardon

Content contributors:
Will Evans

Chelsea Vivash

Publisher:
Laura Waddell

Editors:
Keith Moore

Karen Midgley

Thanks for joining us on the journey of inventive spirit to:

Little Inventors Chief Curator Suzy O'Hara and Creative Producer Jill Bennison.

Our Little Inventors all across the world themselves, with a special mention to Natacha and Rudy Coates, Ruby and Sally O'Hara, Holly and Maya Mataric, Max Evans, Banner Green and Archie Allcroft.

Our amazing Magnificent Makers worldwide!

Cultural Spring, Arts Council England, The Natural Sciences and Engineering Council of Canada, the Victoria and Albert museum, Little Inventors China and Little Inventors UAE, and all of the organisations involved.

Published by Collins
An imprint of HarperCollins Publishers
Westerhill Road, Bishopbriggs,
Glasgow G64 2QT

www.harpercollins.co.uk

The contents of this publication are believed
correct at the time of printing. Nevertheless the
publisher can accept no responsibility for errors
or omissions, changes in the detail given or for any
expense or loss thereby caused.
A catalogue record for this book is available from
the British Library.

Printed and bound in China by
RR Donnelley APS Co Ltd

ISBN 978-0-00-830615-1

10 9 8 7 6 5 4 3 2 1

" Every child is an
artist. The problem
is how to remain
an artist once
we grow up. "

Picasso